ECOSYSTEMS

SR

SALLY RIDE
SCIENCE

Our Changing Climate
ECOSYSTEMS

Contents

[1] ECOSYSTEMS ON THE EDGE

Earth's climate is changing. It's getting warmer. As our planet's air, water, and soil grow warmer, its wildlife hunts for new habitats when old ones become unlivable. Countless species are on the move, from tiny corals to towering pine trees, from slow-flying butterflies to fast-swimming whales, and from curious penguins to fearless polar bears. They're in search of cooler waters, greener meadows, or higher ground. Nature is waving a red flag. Ecosystems are changing!

[2] CHANGE IN THE AIR

Is climate just a fancy word for weather? No. Climate is related to weather, but it's not the same. Weather is what you see when you look out the window. Climate in your hometown is the average weather you can expect where you live. But you can also talk about the climate of a country or a continent or the whole planet.

Temperature's Rising

Earth is getting warmer. That means climates around the world are changing. They're not all changing at the same rate or in the same way—but they're all changing. And that's affecting everything on our planet in one way or another.

Light of Our Lives

That big yellow ball in the sky, the Sun, powers our climate. The Sun constantly emits energy in all directions. Fortunately, a small part of it falls on Earth. Sunlight provides the light and heat that we depend on to live.

Warming Our World

The sunlight that strikes our oceans and land is absorbed at the surface and warms the planet. The warm surface then tries to cool off by radiating the heat back toward space. If this heat could make it out through the atmosphere as easily as the sunlight makes it in, our planet would be much colder than it is. But not so fast! A few gases in the atmosphere—the greenhouse gases—absorb some of the heat before it escapes into space. They trap the heat and make our planet warmer than it otherwise would be. Yes, this is the greenhouse effect (right).

Only 1 Percent

Not all gases are greenhouse gases. In fact, about 99 percent of our air is made of gases that are *not*—oxygen and nitrogen! But without the other 1 percent, there would be no greenhouse effect on our planet. You might think that would be a good thing. Think again.

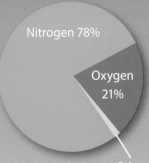

Earth's atmosphere is mostly nitrogen and oxygen.

Nitrogen 78%

Oxygen 21%

Other gases 1%
(including water, carbon dioxide, ozone, and methane)

Meet the Greenhouse Gases

The most important greenhouse gases are water vapor, carbon dioxide, and methane. They're nothing new. They were floating in Earth's air long before there were people on the planet. And though they're only a tiny percentage of our air, those few molecules provide a greenhouse effect that warms Earth. If there were no water vapor or carbon dioxide in our air, Earth would be about 33°C (59°F) colder! Our planet would be one big ice ball.

Okay, That's Enough

If the greenhouse gases in our air keep Earth from freezing, what's wrong with adding more of them? Those gases that we're sending into the air are causing even more warming. And that's affecting our whole planet.

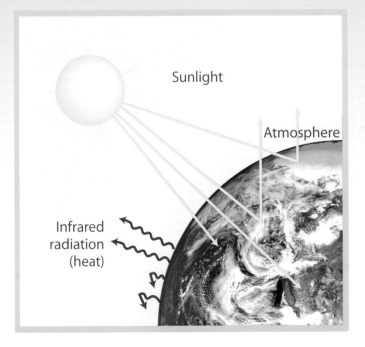

Sunlight

Atmosphere

Infrared radiation (heat)

As carbon dioxide and other greenhouse gases from car and factory fumes build up in the air, our planet is getting warmer and warmer. Uh-oh. Greenhouse overload.

7

Back to the Future

Climate change is nothing new. Over Earth's long history there have been cooler times, like the ice age that happened 15,000 years ago. And there have been warmer times, like the tropical dinosaur days that ended 65 million years ago. In the past, these climate changes were usually triggered by natural shifts in the Earth-Sun orbit, or by changes in the amount of sunlight reaching Earth. That's not the case today.

Uh-Oh. It's Us!

This time, climate change is different. Humans are the cause. How did we do that? We've changed the atmosphere . . . much faster than it's ever been changed before. Many of the things we do—driving cars, flying in planes, lighting our cities, making things in factories—add greenhouse gases to the air. And we're adding lots of them.

How Do They Know?

Aloha, CO_2

Before 1958, no one knew how much carbon dioxide was in the atmosphere. That year, a young scientist named Charles Keeling set up a monitoring station near the top of Mauna Loa, the largest volcano in Hawaii, to find out. He measured the amount of carbon dioxide in the air continuously for many years. His measurements were used to create one of the most famous graphs in science.

The Keeling Curve (right) shows that the amount of carbon dioxide in the air has gone up every year. When the measurements started, there were about 315 molecules of carbon dioxide out of every 1 million molecules of air—or parts per million (ppm). Today, there are around 400 ppm of carbon dioxide! This is a huge increase in a short period.

315 ppm
1958

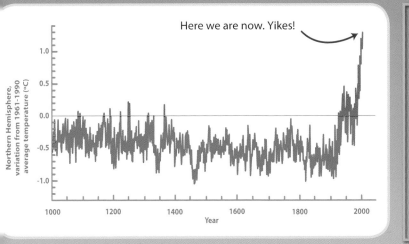

Here we are now. Yikes!

Northern Hemisphere, variation from 1961-1990 average temperature (°C)

Year

Compared to daily shifts in weather, climate change is subtle and hard to measure. It took scientists years to be sure it was real. But it is. In the last century, Earth's climate has warmed about 0.8°C (1.5°F). That may not sound like much, but it's the fastest our planet's global average temperature has changed in 1,000 years.

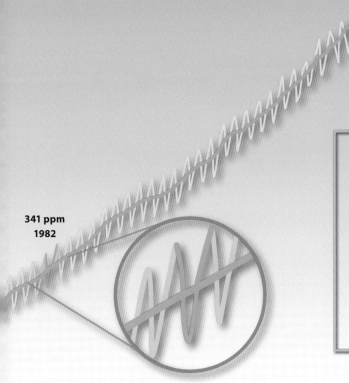

397 ppm
2013

341 ppm
1982

What causes those squiggles? Plants! The graph goes down in the spring, when plants in the Northern Hemisphere grow and suck in carbon dioxide as part of photosynthesis. It goes back up in the fall, when leaves drop from the trees and many plants go dormant. The squiggles show Earth "breathing." But Earth's breathing is the opposite of ours—it inhales carbon dioxide and exhales oxygen.

Warming Signs

So it's getting warmer. What's the big deal? Well, scientists have already measured many changes all around the planet. The oceans are warmer. Glaciers on mountains and ice caps are shriveling. There's more rain in the northeastern U.S., and storms are more intense. There's less rain in the parched southwestern U.S. And ecosystems everywhere are changing.

ON THE MOVE

From the tropical rainforests and hot, dry deserts to the frozen poles and salty oceans, ecosystems around the world are feeling the heat.

Beat the Heat

What do you do when it gets too warm? You look for a place to cool down. That's what plants and animals all over the world are doing as Earth's air, soil, and water heat up. They're flying, swimming, crawling, floating, scurrying, or hitching a ride to places where it's still cool enough to live. Or they're moving into areas that were once too cold for them—but are now warm enough. All over our planet, critters are popping up in places where they've never been before.

A rainforest is just one example of an ecosystem that is feeling the effects of a warming climate.

What's the Big Idea?

Headed Your Way

As temperatures rise in lower latitudes around the globe, animals and plants are expanding into new territories where it's cooler. Earth's creatures are migrating to higher latitudes—an average of 6 kilometers (4 miles) each decade. And some, like the red fox, have already moved north as much as 966 kilometers (600 miles)!

Being There

Ecosystems 101

What's an ecosystem? It's all the living things (plants, animals, and microbes) that live in the same place at the same time, plus all the things that aren't alive there (air, water, soil, sunlight, and so on). But that's only the beginning. It takes energy to stay alive, so each creature must find food. Food chains form. And as microbes, animals, and plants go about the business of life, eating, breathing, and growing, cycles form. The precious molecules of life—such as water, oxygen, and carbon dioxide—pass through living things to the soil, water, or air, and back again and again.

Breaking the Links

Even a small change in temperature or rainfall can have a big effect on plants and animals. Some plants might not survive. Then the animals that feed on those plants would find it harder to get food. An effect on one species always affects others. And when something disrupts one *part* of an ecosystem, it affects the *whole* ecosystem.

An Arctic fox's thick coat, furry feet, and fluffy tail help it to stay warm in the frigid Arctic.

Out-Foxed

The Arctic fox is built for the snow and cold. Its white winter coat blends into the snowy surroundings. Its thick fur keeps it warm in freezing weather. Its larger cousin, the red fox, is made for warmer places. In the past, red foxes and Arctic foxes never lived together. But this is changing. As the Arctic grows warmer and there is less snow and ice, red foxes have started to move in. In a fight between the two, the wily red fox would win. So as red foxes move north, the little white foxes are being pushed into a smaller part of their range, closer to the North Pole.

New Fly Zone

Red-breasted robins? The Inuit people of Banks Island, Canada, were shocked to see these birds. The Inuit have lived on this remote Arctic island for generations. They know the plants and animals in their tundra ecosystem—geese and loons, mosses and low shrubs, musk ox and caribou. Robins are common south of the treeless tundra. Each spring they build their nests and raise their young in forests throughout Canada. But warming trends are drawing them north into the tundra.

Robins are so rare in the Arctic, the Inuit don't even have a word for them.

How Do They Know?

The 866

That's the number of studies recently analyzed by a biologist to find out what's happening to ecosystems worldwide. Ecologists have trekked through forests and waded through streams recording the changing behavior of wildlife. Marine biologists have sailed the seas collecting water samples and measuring the changing biology and chemistry of seawater. Biologists have investigated genetic changes in insects, birds, and plants, looking for recent adaptations. What have they learned? Climate warming is disrupting ecosystems on every continent and in every ocean.

4 U 2 Do

Eco-Drawing

Look up, down, and around. The ecosystem you live in is all around you. Observe a small part of it—under a rock, up a tree, in the schoolyard. Cut a square out of the middle of a sheet of paper, leaving a one-inch border. Use this "eco-frame" to focus on one area. List the living and nonliving parts of your mini-ecosystem. You can also list the microscopic organisms you think are probably there. Watch what the living things do. Then make a drawing that shows some of the links between the living things and some of the links between the living and nonliving things.

Movin' on Up

In a warming world, more and more plants and animals are marching up mountainsides looking for cooler ground. How fast are they moving? More than 6 meters (20 feet) higher in elevation each decade!

Pika's Peak?

If you hurry, you may still be able to hear the squeaks and whistles of pikas talking to each other. Pikas live on the rocky slopes near mountaintops in the American West. The climate is so chilly, pikas wear a thick fur coat—even in summer. Lately, scientists have noticed that pikas are disappearing from places where they once were common. Why? It's getting too warm. The poor pikas have no way to cool off. Temperatures above 31°C (88°F)—even for just an hour—can kill them. What's a pika to do? Keep scurrying up the mountain to stay cool.

The American pika is the smallest member of the rabbit family. When it curls up to sleep, it's about the size of a tennis ball.

End of the Line

Could some ecosystems disappear? You bet. And mountaintop habitats may be among the first to go. Warmer temperatures are forcing some species to move their mountain home higher and higher. Eventually, some may be stranded at the top. Along the way, some species may adapt. But scientists wonder, will there be time? Stay tuned.

New Roots

Birds fly. Fish swim. Rodents run. How do plants move? They let their seeds do the traveling. Seeds blow in the wind and wash away in the rain. Animals help, too. When they chomp on fruit and then defecate, seeds are planted in new places. They are scattered far and wide. Seeds take root if the temperature, soil, and rainfall are just right. Over time, the range of the plant will slowly change. It might die out in the southern part of its range and start to grow farther north.

Not so Sweet

Scientists have their eye on sugar maple trees in New England. The sap from these trees makes the maple syrup you pour on pancakes. But lately, New England has been getting a bit too warm for sugar maples to stay comfy. Sap flows best when temperatures rise above freezing during the day and fall back below freezing at night. This freeze-thaw cycle makes the pressure inside maples just right for sap to flow. Today, farmers are tapping their trees a month earlier than their parents, grandparents, and great-grandparents did. Over time, the trees will migrate north to Canada and largely die out in New England.

Sugar maples won't gallop north one day like a herd of wild horses. But the forest ecosystem will change little by little as the maples disappear.

Custom-Made

Over eons, each species on Earth has adapted to its ecosystem. Its body and behavior are perfectly suited to the conditions and other creatures where it lives. Take a cactus. Cactus doesn't grow at the North Pole. It's made for the hot, dry desert. Its thick stem stores water. Its roots spread under the ground to soak up precious drops of water. Its leaves have evolved into sharp spines. A narrow spine has less surface area than a broad leaf, so it loses less water. Clever.

Species are considered extinct if no one sees them for 50 years. So there's still time. Maybe you'll spot a golden toad someday.

Missing: Last Seen—1989

This could be as close as you'll get to a golden toad. Some forest ecologists think this tiny toad was the first species to go extinct because of climate change. The golden toad made its home in the dripping wet cloud forests of Costa Rica. The toad breathes through its skin and needs moist conditions to survive. But warmer air has pushed the clouds higher up the mountain. The forests are drier—not good for the little toads. Over a few years, the golden toad simply disappeared.

Some Like It Hot, But . . .

The curled horns of desert bighorn sheep are an amazing sight to see. These animals are adapted to the rugged canyons and rocky mountains of the blazing-hot Southwest. But in places such as Anza-Borrego Desert State Park in California, temperatures are getting even hotter—the desert has warmed 1°C (1.8°F) since 1901. And the scarce rainfall has dropped as much as 20 percent. Sparse grasses and shrubs are even sparser. And rare freshwater springs are drying up. As food and water disappear, the sheep are having a hard time.

Most of the bighorn sheep in California's desert mountains have disappeared.

Camille Parmesan

Population Biologist
University of Texas at Austin

There are no limits to what a curious mind can do. "Don't be afraid of looking for answers to big questions," Camille Parmesan says. That's how Camille lives her life, and it's one reason she's become an expert on the impact of climate change on plants and animals. Twenty years ago, people weren't sure how climate change affects ecosystems.

Camille decided to find out. She studied one type of butterfly, Edith's checkerspot butterfly, to see how it was responding. "Populations go extinct during extreme climate years," Camille explains. After years of research—including hikes all over the western U.S. to record the condition of the butterflies in their natural habitat—here's what she learned. The butterfly's range has shifted north by dozens of miles, and populations in the warmer southern range are going extinct at a high rate.

After that, Camille analyzed hundreds of studies (866!), and thousands of species from around the world. She uncovered many of the patterns we know today—species moving toward the poles and higher elevations, flowers blooming earlier in spring, and so on. "Documenting the impacts is essential to convince people that they have to take action on climate change," she says.

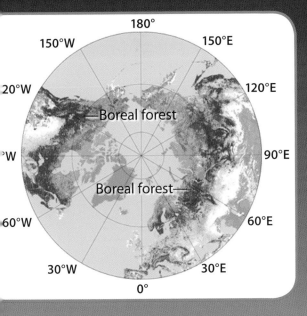

180°

150°W   150°E

20°W   120°E

—Boreal forest

°W   90°E

Boreal forest—

60°W   60°E

30°W   30°E

0°

A Damaged Sink

The boreal forest is like a giant necklace of evergreen trees around the Northern Hemisphere. It covers almost 15 percent of Earth. That's a lot of trees and a huge carbon sink. Shwoosh! The trees suck in a whole lot of carbon dioxide from the air. But warmer temperatures are drying out the forest soil. The trees aren't soaking up much carbon dioxide— they've almost stopped growing. You know what that means—more carbon dioxide will stay in the air.

Did You Say Sink?

Trees take in carbon dioxide gas from the air and use it in photosynthesis to make their own food. Eventually, the carbon in the gas molecule becomes part of the tree—its leaves, stems, roots, and the wood in its trunk. The tree is a carbon sink. The tree holds carbon like a sink holds water. But it doesn't hold it forever. When leaves fall or the tree dies, they are eaten by other living things. As part of this digestion, carbon dioxide is released back into the air.

How Do They Know?

Eyes in the Skies

Ecologists use satellites in space to get a global view of ecosystems. Satellites carry cameras and sensors that measure light reflected from grass, rocks, water, and everything else. Each type of plant reflects light in its own way. This is called its spectral signature. Since each ecosystem has characteristic types of plants, scientists can easily spot different ones in satellite images. For example, lichens and low plants grow in the tundra. They don't look the same as forests of spruce trees in the boreal forest. Over time, satellite images can show changes in ecosystems.

[4] WARMING WATERS

Even Earth's largest ecosystems, the oceans, are warming up. Many rivers and lakes are also warmer now than in the past. Maybe this doesn't sound like such a bad thing. But more than 70 percent of our planet's surface is covered in water. So warming waters mean big changes all over the globe.

At a Snail's Pace

A few years ago, marine biologists started to see something unusual along the coast of Northern California—a tiny tube snail, far from home. The snail is common along the warmer rocky seashore in Southern California. But up north? Just as on land, aquatic creatures move as climate changes. So aquatic ecosystems are changing, too.

Moving In, Moving Out

The rocky tide pools of California's Monterey Bay are a bustling ecosystem, crowded with barnacles, crabs, sea anemones, and seaweed. Each creature is perfectly adapted to a life spent half in water at high tide and half out of water at low tide. Over the past few decades, water in the bay has warmed 1°C (1.8°F). Did this affect life in the tide pools? To find out, researchers walked part of the bay, counting every creature living there, exactly as marine ecologists had done in the 1930s. What did they find? Big changes in tide-pool tenants. In all, six species adapted to cooler conditions had disappeared. Meanwhile, 10 species suited to warmer waters had moved in, including the tiny tube snail from down south. It now covers the rocks!

The brook trout—Pennsylvania's state fish—has shifted its range to higher, more remote streams.

Gone Fishin'

Expecting a trout on the end of your line? Warming waters could change that. Many freshwater fish are swimming to cooler waters. Trout and salmon need cold waters. These swift swimmers need lots of oxygen to power their sleek bodies through the water. Oxygen dissolves more easily in cold water than in warm water, so cold water holds more of it. If the fish can't migrate to cooler waters, their ranges will shrink. But some fish—such as bass, catfish, and carp—thrive in warmer waters, so their ranges are expanding.

Good Day, Sunshine

Even though our Sun is 150 million kilometers (93 million miles) away, it powers all life on Earth, including ours. Day in and day out, energy from the Sun lights and warms our planet. Some of this energy enters ecosystems when plants on land and phytoplankton in water perform photosynthesis—they use the energy in sunlight to turn carbon dioxide and water into food. The food is carbon-carrying sugar molecules packed into the cells of phytoplankton floating in the ocean, or in the berries on a blueberry bush, for example. One way or another, this is the food that fuels nearly all life on Earth. And it all starts with a sunny day.

The Faces of Our Planet

We share Earth with countless other living things—from microscopic plankton to massive polar bears. They depend on us—just as we depend on them—to keep our planet healthy. As Earth grows warmer, many species around the world are feeling the heat. Some will be able to adapt and some will not. These are some of their faces.

Polar bear

Robin eggs

White dogwood

Red squirrel

Monarch butterfly

Andean flamingo

Lake Tanganyika snail

Flightless cormorant

Diatom

Siberian crane

Natterjack toad

Tibetan antelope

Snow monkey

Red ruffed lemur

Koala

Quiver tree

Chinstrap penguin

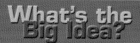

Oh, By the Way

The oxygen in our air? It comes from plants, too! They just happen to produce oxygen during photosynthesis, and it floats up into the atmosphere. If there were no phytoplankton or trees on Earth, there would be no oxygen to breathe. Thanks, all you green guys!

4 U 2 Do

Clues Will Do

Can you tell the story of a food web—with *no* words and *no* pictures of plants or animals? Try to tell the story to a friend, in drawings of *only the clues and evidence* of the web. For example, if you want to explain that a hawk ate a mouse, draw mouse tracks that suddenly stop, and draw a feather next to them. Draw one piece of evidence for each link in the web.

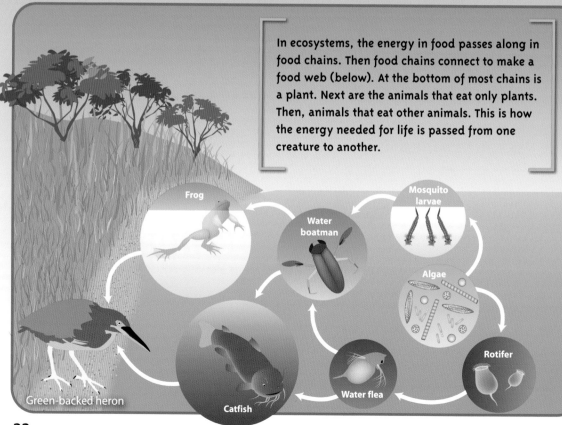

In ecosystems, the energy in food passes along in food chains. Then food chains connect to make a food web (below). At the bottom of most chains is a plant. Next are the animals that eat only plants. Then, animals that eat other animals. This is how the energy needed for life is passed from one creature to another.

Frog

Mosquito larvae

Water boatman

Algae

Rotifer

Water flea

Catfish

Green-backed heron

22

Breaking the Chains

Lake Washington near Seattle used to be a vibrant ecosystem supporting algae, zooplankton, and salmon. Each spring, green algae bloomed at the same time *Daphnia*—a zooplankton—populations peaked. *Daphnia* gobbled up the algae. Then young sockeye salmon (right) chowed down on the *Daphnia*. But fish biologists have noticed that this cycle of life is out of whack. The temperatures in the upper waters of the lake have warmed about 1.6°C (2.5°F). Algae are blooming earlier in spring. *Daphnia* are not. Fewer *Daphnia* are surviving, so there's less food for the young salmon. Some salmon are struggling to make their way upstream to spawn.

How Do They Know?

Thermometer in Space

Every day, satellites in space measure the temperatures of Earth's oceans. Information is beamed back to Earth where computers turn it into maps of sea surface temperature. What have we learned? The average temperature of the oceans has risen 0.04°C (0.07°F) in just the past 50 years. That may not sound like much, but remember, it's averaged globally. So some parts of the ocean are warmer than others. In tropical waters, average temperatures have risen as much as 1.7°C (3.7°F), and many coral reefs are struggling to survive. In the Arctic, waters have warmed even more and sea ice is melting.

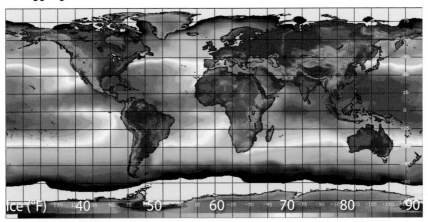

Ice (°F)   40   50   60   70   80   90

Undersea Rainforests

Coral reefs are found in warm tropical waters like a belt around the equator. These ecosystems are called underwater rainforests because they're teeming with life. Coral reefs furnish food and shelter for everything from sea urchins, sea snakes, sea horses, and sea turtles to stingrays, eels, octopuses, and dolphins—not to mention all the microbes.

Tiny but Mighty

Australia's Great Barrier Reef is more than 1,610 kilometers (1,000 miles) long! But the building blocks of the reef—coral polyps—are tiny animals no bigger than your thumbnail. Each little coral builds a calcium carbonate skeleton around its soft body to protect it. The corals are the living outer layer of the reef. Generation after generation of corals lays down new layers of rocky skeleton on top of old layers. Over time, a reef is made—no architect needed.

Coral S.O.S.

Many corals are struggling to beat the heat. When waters around them are too warm for too long, corals expel the colorful algae that live inside them and turn white. This is called coral bleaching. The coral animal provides protection for the algae, and through photosynthesis, algae make the food that feeds the coral. So without their houseguests, corals eventually starve. Bleaching can ripple through the ecosystem. Dead corals erode and reef fish lose their hideouts and feeding grounds, turning the coral reef into a ghost town.

Many corals are living in waters where a temperature increase of just 1°C (1.8°F) would be deadly.

Being There

Coral Comeback

Staghorn and elkhorn corals once thrived in South Florida and throughout the Caribbean. Today their colorful colonies are dwindling—so much so that they were placed on the threatened species list. The situation isn't hopeless, though. Scientists and conservationists are working to restore the battered reefs. Five kilometers (3 miles) off the coast of Key Largo, Florida, a coral nursery—fragments of different species of coral glued on cement blocks—is being nurtured. By figuring out which corals thrive, scientists may be able to restore the damaged reefs. Come back, little coral!

With their thick, flat stalks—which look a bit like elk antlers—these corals can easily be identified underwater.

Air Today, Oceans Tomorrow

About one-half of the carbon dioxide we pump into the air ends up in the oceans. It dissolves in the seawater, slowly changing its chemistry. As carbon dioxide dissolves, it combines with water molecules to form carbonic acid. Uh-oh. This is making the oceans more acidic. In fact, the oceans are more acidic than they've been in millions of years!

No Bones About It

Another chemical in seawater, called calcium carbonate, neutralizes the acid buildup. Scientists worry that someday the level of calcium carbonate could drop dangerously low. Calcium carbonate is the favorite building material of countless sea creatures, from corals to crabs to phytoplankton. But if carbonate levels drop too low, these animals won't be able to build their protective shells or skeletons. It would be like trying to build a brick house with only bricks and no mortar.

Attack of the Ocean Squishies

The sea squirt won't win any beauty contests. Folks say they resemble soggy scrambled eggs. Lately, sea squirts are causing trouble around the world, and it's not their looks creating the problem. As sea temperatures rise, sea squirts are moving into new habitats and taking over. They carpet the seafloor in thick, spongy mats, muscling out native bottom dwellers like clams, scallops, and worms. Georges Bank, off the coast of Massachusetts, is one of the Atlantic's most important fishing areas. It's under siege by a massive sea squirt invasion.

It's All About Light

Earth is shaped like a giant beach ball—a sphere. So different parts of our planet receive different amounts of sunlight. The equator receives the most—a direct hit of sunshine. The poles receive the least—grazing rays of sunlight. These differences determine worldwide patterns of weather and climate. In turn, they help determine which plants and animals live where. Presto—all the world's ecosystems! See the patterns? Similar ecosystems roll out across each continent and ocean in about the same places.

Land Ecosystems

Chaparral	Polar ice and high-mountain ice	Temperate grassland
Coniferous forest	Savanna	Tropical forest
Desert	Temperate deciduous forest	Tundra

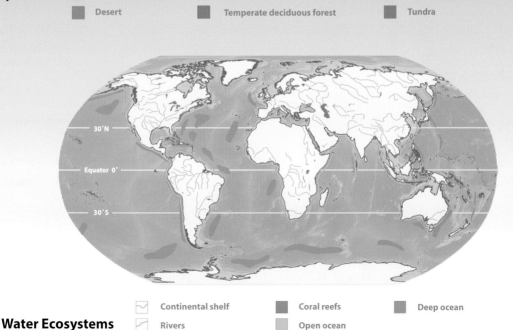

Water Ecosystems

Continental shelf	Coral reefs	Deep ocean
Rivers	Open ocean	
Area of high and low tide	Lakes and wetlands	

MELTING ICE

5

What happens when you leave ice out on a warm day? What happens when you warm up the air around Earth's thick ice caps and glaciers? They melt, too!

Melt-o-Rama

Icebergs, ice sheets, and glaciers. Ice covers about 10 percent of our planet's surface. Ice is *really* bright. Ice reflects about 90 percent of the sunlight striking it. The light bounces right back out into space, so it doesn't warm our planet. What happens when ice melts? The land or water under it is exposed. These darker surfaces absorb sunlight instead of reflecting it back to space the way ice does. So the sunlight warms the water or land—and even more ice melts. And then more . . .

Top of the World

Over the last century, temperatures in the Arctic have increased at twice the global rate. In some places, the Arctic is as much as 4°C (7°F) warmer than it was only 50 years ago. This makes it harder and harder for sea ice to form. Less and less sea ice is affecting ecosystems all across the Arctic.

1979

2012

On Thin Ice

As Arctic sea ice grows smaller and thinner, life gets tough for animals that have depended on it for centuries. Polar bears spend most of the year roaming the ice, hunting for seals that swim in near-shore waters. Then they spend summers on land, living off stored fat. In Canada's Hudson Bay, the ice is breaking up three weeks earlier than it did just 30 years ago, and "ice-up" is happening later in the fall, so the bears have less time to hunt. Polar bear populations are down 22 percent, the bears weigh less, and females are having fewer cubs.

Winners and Losers

Much of the Bering Sea used to freeze each winter. Now some of it doesn't. Today the southeastern Bering Sea is at least 2°C (3.6°F) warmer than it used to be. There's less sea ice. Algae live on the bottom of sea ice. Some of this algae falls to the seafloor. This provides a feast for bottom-dwelling sea life, such as worms, crabs, and flatfish. When seas are warmer, tiny animals called zooplankton thrive. They eat the algae, leaving less of it to drift down to the seafloor. So populations of bottom dwellers are down. But pollock fish like to snack on zooplankton. Their populations are up. Pollock like the change. Crabs? Not so much.

Walruses use their flat noses to plow the seafloor for clam dinners.

Walrus Woes

In the spring, Pacific walruses haul themselves out onto floating sea ice to mate and raise their young. These SUV-sized mammals live together in herds. When they get hungry, they barrel into the water to chow down on shrimp, clams, and amphipods on the seafloor. But as the Bering Sea warms and the sea ice shrinks, walruses are forced to move farther north. Sometimes the water is too deep for walruses to reach the bottom—so they go hungry. Lately, scientists are seeing fewer pups.

The Skinny on Gray Whales

Sightings from Alaska to Mexico are all the same. Gray whales are too skinny! Scientists are seeing scrawny flanks, bony shoulders, and flat humps. Why aren't they seeing plump, round bodies with lots of blubber under the skin? Slim pickings at suppertime. Before they make their long migration south, whales bulk up on small crustaceans called amphipods. Amphipods live on the seafloor and survive by eating algae falling from the ice above. But as Arctic ice melts, the whales' favorite food is off the menu. So far, the scrawny whales are holding their own by switching to other food. Whew! The gray whale is learning to adapt.

Gray whales spend the winter in Baja California, Mexico, breeding and raising their calves before swimming north to Alaska each summer.

Russia

Chukchi Sea

Bering Sea

Alaska (United States)

Canada

Pacific Ocean

United Sta

Mex

—— Gray whale migration

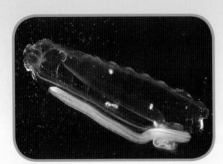

Salps are see-through. They eat by pumping seawater through their bodies and filtering out phytoplankton, like an old-time gold miner panning for gold.

Here Come the Salps

While the krill population around Antarctica is down, the salp population is up. What's a salp? A small, barrel-shaped sea creature. Sometimes salps link together and form long, stringy colonies 6 meters (20 feet) long. As the Southern Ocean warms, salps are floating farther south. They spend their nights feasting on phytoplankton. The problem is they eat too much. So there are no leftovers for the zooplankton . . . which feed the krill . . . which feed the whales, seals, and penguins. So now waters once full of krill are full of salps—but *they're* not part of this food chain.

Disappearing Penguins

Ice is melting at the South Pole, too. The air has warmed—nearly 3°C (5°F)! The Southern Ocean is warmer. No wonder ice shelves and glaciers are melting in Antarctica. Emperor and Adélie penguins like to float on rafts of sea ice. But with less sea ice each year, penguin populations have taken a nosedive. Why? Because of broken links in the food chain. Algae live under the sea ice. Shrimp-like animals called krill eat the algae. Penguins lunch on the krill. Less sea ice means there's less food for the penguins.

Glacier-*less* National Park

Glacier National Park may need a new name soon. In 1910, when the park was opened, there were 150 glaciers studding the landscape. Today, only 27 are left! Scientists predict the last of the park's glaciers will have thawed by 2030.

Being There

Meltdown

Glaciers are disappearing all over the world. In the mountains of India, Bolivia, and Peru, glaciers are like giant watercoolers. More than a billion people rely on the glaciers for fresh water—to drink, to irrigate crops, and to make electricity. But they're melting—faster than anyone thought possible.

Researchers climb high into the Himalayas to measure the melting and movement of glaciers.

6

RISING SEAS

Look out! Seas are on the rise. Warm water takes up more space than cold water. So as the oceans warm, seawater expands and sea level creeps up.

How Do They Know?

Sea Swell

Tide gauges on ocean piers and satellites in space measure the rise and fall of oceans around our planet. They show that seas have swelled about 15 to 20 centimeters (6 to 8 inches) in the last century. About half that rise is from expanding seawater. The other half is from melting ice adding fresh water to the salty oceans.

From Coast to Coast

It isn't just plants and animals that lose their homes when sea levels rise. People do, too! For thousands of years, people have built their villages, towns, and cities near the ocean. Today, more than 600 million people live along the coast—from Cairo to Cape Town and from Los Angeles to London. Rising seas could chase millions around the world to higher ground.

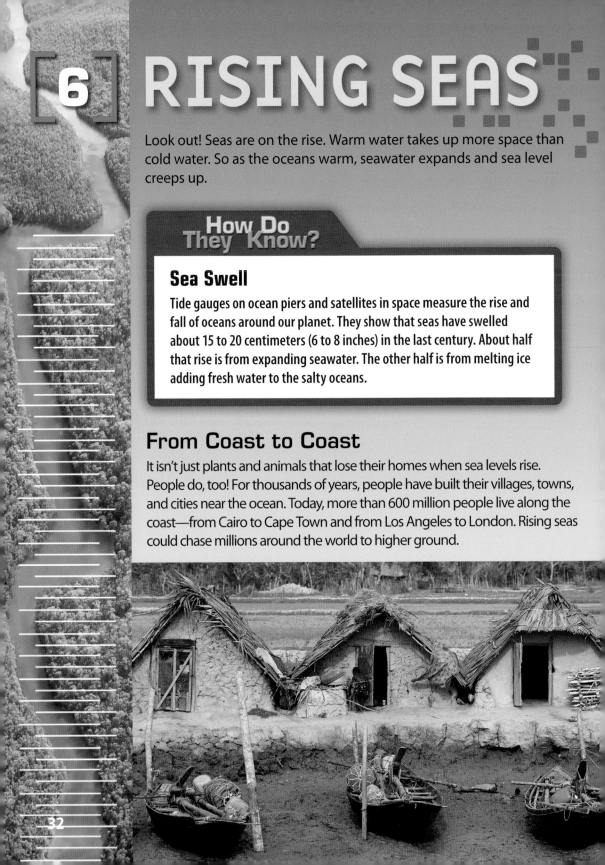

Washing Away

The rising ocean is eating away at many of the world's wetlands. The Mississippi River Delta is the largest wetland habitat in the U.S. But every 30 minutes, it loses an area the size of a football field—it washes into the Gulf of Mexico. Some of this loss is due to rising sea levels, and some to natural sinking of the soil. As the delta disappears, it takes with it a salt-marsh ecosystem crawling with critters. Crayfish and crabs hide under rocks. Pelicans and egrets feast on worms. Rows of alligators bask in the sun. If waters in the gulf keep rising, scientists predict that one-third of these wetlands will be under water by 2050.

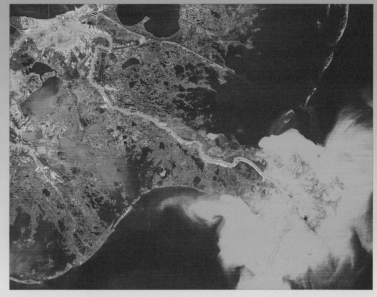

From space, the Mississippi River Delta looks like a giant bird's foot.

4 U 2 Do

The Ecologist in U

Pick an ecosystem that you don't know much about—maybe a coral reef, rainforest, or tundra. How is it being affected by climate change? Pick your favorite creature there and chart its niche in the ecosystem. Where does it live? Does it live by itself or with others? What does it eat? What's the rest of its food chain? How would the ecosystem change if your favorite critter were no longer a part of it?

[7] SEASON CREEP

Many of the things plants and animals do are controlled by environmental cues—temperature, sunlight, time of year. These cues tell plants when to bud and when to unfurl their leaves. They let animals know when to migrate south for the winter or when to wake up from hibernation. So what happens when winters are milder? Spring comes early? Fall frost comes late? Things get confusing!

Hey, Bud

What's with spring? In many places, it's much warmer and it's arriving much earlier. At the poles and in temperate latitudes (between the tropics and the poles), spring starts one week earlier than it did only 20 years ago. Trees are getting cues to bud and bloom earlier. The famous cherry blossoms in Washington, D.C., are bursting open several days earlier than ever before. Weird.

Winter Freak

Rufous hummingbirds spend their summers in the forests of western North America. They nest in colonies—sometimes 20 nests only a few meters apart. But as winter approaches, the tiny birds fly south. They spend the cold winter in Mexico. So winter sightings in the U.S. are rare—or at least they used to be. Lately, people have been spotting the hummingbirds all along the Gulf Coast where winters have warmed. This colorful little bird's winter range has shifted big-time.

Down on the Farm

Farmers are feeling the heat, too. Wheat, corn, and rice just aren't growing like they used to. The air is warmer. The soil is drier. But these grains make up a big part of people's diets. Breads, tortillas, or rice cakes are on dinner tables around the globe. The news isn't all bad. Farmers are adapting. They're planting crops earlier in spring. And they're trying out new crop varieties that can handle the heat.

No-Doze

Bears hibernate in winter. Their metabolism slows down, and they barely breathe, to conserve energy. They sleep during this coldest time of year when food is scarce. So where did those fresh bear tracks come from? With milder winters in the mountains of northern Spain, some female brown bears are skipping hibernation. They're shaking themselves awake (their cubs, too) and roaming the mountainsides. They're finding nuts and berries instead of snow. And the big, brown bears are taking hikers by surprise.

How Do They Know?

Ring Them Trees

Did you know that trees keep climate journals? Their journal entries are the fat and thin rings of new wood laid down each year as the tree grows. Trees respond to temperature, rainfall, and other conditions by growing a little or a lot. Scientists study tree rings to learn about past climate. The number of rings tells them how old the tree is. The width of the rings tells them which years were wet and which were dry. Sometimes rings have scars from fires or floods or pest invasions. If a tree is really old, this climate record can stretch back hundreds or even thousands of years.

What's the Big Idea?

Another . . . -ology

What do ecologists call one of the ways animals, plants, and microbes respond to the climate? Remember the mama bears that didn't hibernate? And the algae that bloomed earlier in spring than the zooplankton? The timing of events in the life of a species is called its phenology.

8

INVASION!

We're not being invaded by aliens from another planet, but by alien plant, animal, and microbe species moving into ecosystems where they don't belong. Living things are on the move. But not all migrations are helpful. Some are harmful—especially when species that carry diseases or run wild invade an ecosystem.

Not Just a Bug Bite

Some insects are being lured into places once too cold for them, like Nairobi, Kenya. Mosquitoes never used to hang out there. Now they're buzzing in this mile-high city. And they're bringing disease with them. The disease? Malaria. Some tropical mosquitoes carry a parasite in their mouths that causes malaria. When a mosquito bites, its saliva lets loose a flood of these freeloaders. The parasites make their way through a person's body, eventually invading her or his red blood cells. The symptoms—burning fever, headache, and diarrhea. Yikes! And, if that weren't enough, now that person is a carrier.

This mosquito has a belly full of blood.

Watch Your Step

Poison ivy is pumped! With more carbon dioxide in the air, the weed is growing like gangbusters—twice as fast as it used to. The vine grows so fast that it chokes out young trees in forest ecosystems. But that's not all. Weed ecologists say that as carbon dioxide in our air continues to rise, the rash-causing oil that poison ivy makes will get even more potent. Ever made the mistake of brushing up against some? Oh, that itchy, itchy rash!

Icky *Ixodes*

A pinhead-sized parasite is becoming a pest in Sweden. It's a tiny tick—*Ixodes ricinus*. Entomologists say the little bloodsucker is popping up on cats and dogs and people in places where it was once rare. Sweden now has fewer cold days in winter and more warm days the rest of the year—perfect conditions for icky *Ixodes*. The tick hides out in the grass and ambushes any passing animal. Like little vampires, ticks need blood to complete their life cycle—from larva to nymph to adult. Warning to all Swedish dogs and cats: Stay home!

Ticks bug pets, but they also carry diseases that make humans sick.

Experts Tell Us — Neil Cobb

Ecologist
Northern Arizona University

After several years of unusually hot, dry weather, something strange happened in Arizona's piñon-juniper woodland. Bark beetles seemed to come out of nowhere, attacking the piñon pines. Neil Cobb was studying another insect pest in piñon pines at the time. "Then the bark beetles started eating my trees," he says. That got his attention.

After almost 10 years of drought, the pines were weakened. Then the explosion of bark beetles finished them off. Why now? "There have been droughts before, but this is a hotter drought," says Neil. Neil suspects global warming is making a bad situation worse.

Piñon pines and junipers used to grow in a 50-50 mix. Now there are mostly junipers. The loss of the piñon pine is being felt by piñon jays, piñon mice, and hundreds of other creatures of this desert ecosystem. Eventually, the range of the piñon pine will shift north from Arizona to Colorado. "We should have some sense of how we're changing the planet. Then we'll have some idea of what we might do to make things better," says Neil.

9

WHAT CAN WE DO?

Ecosystems are often interrupted by roads, buildings, or fields of crops. So people are setting aside wildlife corridors. These corridors connect isolated patches of an ecosystem. If water is scarce in one patch, creatures move through the corridor to reach another patch where there might be some. These corridors also provide a way for birds, insects, plants, and mammals to migrate to cooler places.

Seeing the Forest *for* the Trees

Trees are global-warming fighters? Oh, yes! Trees suck in carbon dioxide, *lots* of it, for photosynthesis—at least one-fourth of the carbon dioxide we put into the air. That's why people around the world are preserving their forests. And they're planting trees to restore forests that have been cut down. No more chop, chop. Plant! Plant!

Experts Tell Us / Rodolfo Dirzo

Tropical Ecologist
Stanford University

Dinnertime in the tropical forests of his native Mexico is the best part of the day for Rodolfo Dirzo. But it's not about what's on his plate for the tropical ecologist. Rodolfo's more interested in the animals and plants of the forest and who's eating (or not eating!) what.

Rodolfo is an expert on the balance between animals and plants in ecosystems. "Many things in nature seem mysterious at first. But they make sense if you understand how they are all connected," Rodolfo says.

Working Together

All over the globe, nature is sending us warning signs. You can feel them in the hot, dry wind. You can hear them in the whistles of the pikas. And you can see them in the bleaching coral reefs, melting glaciers, and drying forests.

But all over the globe, people are paying attention! We are a very clever, creative, and caring species. We are coming up with solutions for keeping our planet healthy. And the solutions are catching on. Switch to clean energies. Recycle. Take care of Earth's precious air, water, and life. You can help, too. Make wise choices about how you use energy, and encourage your friends and family to do the same.

atmosphere (n.) the mixture of gases (nitrogen, oxygen, and traces of others) surrounding Earth, held in place by the force of gravity (pp. 6, 8, 22)

climate (n.) prevailing weather conditions for an ecosystem, including temperature, humidity, wind speed, cloud cover, and rainfall (pp. 6, 13, 27, 35)

ecosystem (n.) all of the living organisms (plant, animal, and microscopic species) in a given area that interact with each other and their surrounding environment (pp. 4, 12, 19, 27, 33)

extinct (adj.) no longer existing, as in the case of a species where all individual organisms are lost (pp. 15, 16)

food chain (n.) the path of food from one living organism to another in an ecosystem, showing who eats whom (pp. 11, 22, 30)

food web (n.) the pattern of interconnecting food chains in an ecosystem. (p. 22)

fossil fuel (n.) a nonrenewable energy resource such as coal, oil, or natural gas that is formed from the compression of plant and animal remains over hundreds of millions of years (p. 9)

greenhouse effect (n.) the warming that occurs when certain gases (greenhouse gases) are present in a planet's atmosphere. Visible light from the Sun penetrates the atmosphere of a planet and heats the ground. The warmed ground then radiates infrared radiation—heat—back toward space. If greenhouse gases are present, they absorb some of that infrared radiation, trapping it and making the planet warmer than it otherwise would be. (pp. 6, 7)

greenhouse gas (n.) a gas such as carbon dioxide, water vapor, or methane that absorbs infrared radiation, or heat. When these gases are present in a planet's atmosphere, they absorb some of the heat trying to escape the planet instead of letting it pass through the atmosphere. The resulting warming is called the greenhouse effect. (pp. 6, 7, 8)

habitat (n.) a place where individual organisms of a particular species live. It provides the types of food, shelter, temperature, and other conditions needed for survival. (pp. 4, 13, 16, 26)

photosynthesis (n.) the process by which plants and other photosynthetic organisms use energy from sunlight to build sugar from carbon dioxide and water. As part of this process, oxygen is released. (pp. 9, 17, 19, 22, 25, 38)

INDEX